Cottonwood Trail

~Schmunk

~Schmunk

~Davis

~Schmunk

~Davis

~Schmunk

~Schmunk

~Schmunk

~Davis

~Schmunk

Cottonwood Trail

Glimpses of Wildness in the Heart of Spartanburg

essays by Thomas Webster
photography by G.R. Davis Jr. & Peter L. Schmunk

Hub City Writers Project
Spartanburg • 2006

ISBN 1-891885-48-0
First printing—October 2006

Front cover and title page photographs—G.R. Davis
Back cover photograph—G.R. Davis
Cover & book design—Mark Olencki
Project editor—Betsy Wakefield Teter
Copy editing—Carol Bradof, Tom Johnson, & Michael Smith
Scanning—Mark Olencki & Christina Smith
Printed in China

Hub City Writers Project
Post Office Box 8421
Spartanburg, South Carolina 29305
(864) 577-9349 • fax (864) 577-0188 • www.hubcity.org

Publication of this book is funded in part by the Arts Partnership of Greater Spartanburg and its donors, the City and County of Spartanburg, and the South Carolina Arts Commission, which receives support from the National Endowment for the Arts.

TABLE OF CONTENTS

PREFACE

by John Lane

SINCE THAT FIRST STORIED 1995 MEETING IN Morgan Square Coffee of its three founders, the Hub City Writers Project has always been a "place-based" literary organization. In our first book, *Hub City Anthology*, three of the twelve essays directly explored the relationship between nature and culture in our piedmont community.

Other Hub City books soon followed that celebrated the landscape of Spartanburg County—*Family Trees: The Peach Culture of the Piedmont*, *The Lawson's Fork: Headwaters to Confluence*, *The Seasons of Harold Hatcher*, and *Noble Trees*.

Even the historical chronicle of Spartanburg's cotton mill culture, *Textile Town*, begins with an explanation of the geology of our area and why the abundant shoals on our Upcountry rivers became the "bedrock" of the industrialization of the piedmont. History, for Hub City, has always included natural history.

Anyone familiar with Hub City knows that our work is not limited to publishing books. In 2000, at Hub City's behest, renowned nature writer and National Book Award winner Barry Lopez visited Spartanburg, walked the Cottonwood Trail, and then read to a large crowd at the public library. He reminded us that Lawson's Fork is not "impaired," as DHEC has classified it; instead it is the community's relationship to the creek that is impaired. The Lawson's Fork is not a "drainage ditch" for Spartanburg's economic growth. Instead, it's a living stream. We needed to celebrate it. No telling what would happen if we did.

Later that year, with Lopez's insight in mind, Hub City celebrated the creek and our relationship with it with the publication of *The Lawson's Fork*. For four consecutive days a festival moved down the length of the creek, from Inman to Glendale. During that week thousands of people visited the creek, many for the first time. They enjoyed music, took canoe rides, saw environmental art, and walked along the Cottonwood Trail with our homegrown naturalist supreme, Rudy Mancke.

Along the way we carried an urn shaped of local clay, full to the brim with creek water. Children from all over the watershed gathered the waters at the festival's beginning and returned the water to the creek's ceaseless flow four days later, blessed on its journey by an Episcopal priest and a Cherokee shaman.

Those who were there on the final day of the festival at the ball fields below Spartanburg High will also remember that another delegate was in attendance. A huge red-tail hawk sat on one of the light poles and listened while the creek was praised and welcomed into our human community. The hawk sat and listened as the gospel choir sang "The Water is Wide" and stayed put while the Spartanburg Philharmonic played. Its cries punctuated the ceremonies occasionally like applause. I took it as approval.

So *Cottonwood Trail* is the latest in a long line of Hub City projects meant to reveal this place to its people. In what follows Thomas Webster explores the eastside trail for a year and writes a series of brief, poetic essays about what he sees when he walks this 85-acre patch of "nearby nature."

G.R. Davis and Peter Schmunk do the same with cameras—they visit the trail through the cycle of the seasons and come away with a series of images that reveal a place of magic, diversity, and hope. Their photos are a treasure for those who have always known that although the Carolina piedmont is not Yellowstone, it has its beauty and transcendence. You just need to know where and how to look.

Enjoy this field trip into the compelling world of the Cottonwood Trail, what we're calling "glimpses of wildness in the heart of Spartanburg." Here we're showcasing one of Spartanburg's "beauty spots." We're also introducing you to a new Hub City writer and two new photographers. Add these voices and visions to the natural history of this place.

—John Lane, a founder and board member of the Hub City Writers Project, is the author of three books of nature writing, Chattooga: Descending into the Myth of Deliverance River, Waist Deep in Black Water *and* Weed Time.

Introduction

by Thomas Webster

It is my favorite time of the morning on the Cottonwood Trail, that hour when every strand of spider's silk becomes a silver banner. A woodpecker is taking some half-hearted taps at a nearby trunk. This morning I have seen a spider catch a grasshopper, and shortly afterwards, a finch plucked a neighboring spider from its perch.

By the calendar of these woods, this is the time of jewelweed. In this time, you might cross paths with the lacquered box of a turtle as it considers a burrow or witness the last independent days of a robin before it joins a foraging flock. During these days pokeweed bows over and presents its berries to the earth. Waning is the hour of the bumblebee. The muscadines are becoming a too-sweet memory, and the fallen flowers of trumpet vine are all but forgotten. Coming are the days of bluster and ice. The reign of the mosquito is drawing to a close.

Bald-faced hornets created their own tree-house in the preserve. *~Davis*

I sit by the tree that once held my boyhood tree house, the evidence of which has eroded away from everything but my memory. To get here I followed our old path, a route forged before the Cottonwood Trail was established. For a significant portion of my youth, I came to these woods nearly every day. I sometimes wonder what it was that we sought here, my friends and I, for I can no longer recall. Just being boys, I suppose.

Nothing as convenient as that discerns what I now seek. What keeps me coming back to this place? After all, a wife, three children, and a career away from my youth, I have given up most of my former pursuits. I no longer haunt the skating rink, and, although I wish I hadn't, I have given up dedicating entire summer afternoons to touring the neighborhood by bicycle. Those things and many more I have left behind. But not rambling around these woods. Something on these eighty-five acres of Spartanburg's eastside keeps me coming back.

Part of it, I suppose, is the satisfaction inherent in being so familiar with a place, of knowing it. In these woods treasures mark every season, and you learn to keep an eye out for them, unwilling to let them pass unnoticed. Soon the falling leaves will unveil a hornets' nest that is further along the path. The American beech will refuse to give up its leaves, and though they will fade to the color of old parchment, they will not fall until new growth pushes them aside in the spring. The ebony jewelwing will flit out of the wetland, and my daughters will stand as still as reeds until dragonflies light upon their upraised fingertips.

Even as I sit here, the breeze suggests the winter that it will usher in, and I know that, in time, the ground will be thick with cottonwood "cotton" instead of leaves. While it's true that poison ivy will again teem from the margins of the trail, it is also true that honeysuckle will join it.

The surprises, too, keep me coming back. Like the day I was walking the path and discovered that honeybees had set up shop in a knothole at the top of a Tree-of-Heaven. Or another day, months later, when I came to the same spot and found that a possum had likewise discovered the hive. Bits of honeycomb littered the earth at the base of the tree, testifying to the exuberance of the feast. I used to chance upon a squirrel that was entirely white, and sometimes still, I will round a corner and find myself presented with a great blue heron as it gazes into the waters of the creek. For two seasons black morels, as unexpected as a comet, appeared upon a single hummock. And not far from that spot I sometimes encounter the sleek form of a river otter before it vanishes and leaves me staring at its wake.

I come here, too, I think, because places like this are increasingly rare. The Edward M. Griffin Preserve (on which the Cottonwood Trail is forged) is a miniscule portion of the 12,075 acres that comprise the city of Spartanburg. Yet only another thirty-three city acres are likewise protected. In a county that is developing more acres per person than any other in South Carolina, these eighty-five acres remain and *will* remain undeveloped.

This place matters because it is full of things that you rarely see out in those unprotected (and often built upon) acres. In these woods my approach has triggered a fusillade of leopard frogs. As my children collected the empty shells of freshwater mussels, I have marveled at the petroglyph of a raccoon's paws. In these woods I have startled a doe, and my little girls and I have given laughing chase to a retreating turkey. I have witnessed a chipmunk tumbling into the water, and I have seen that same water, as milky blue as a jellyfish, washing over a sunbeam caught in its depths.

In these eighty-five acres, on these three and a half miles of trail, exists more mystery than even the most generous lifetime will afford me to wonder over. And we should never imagine that it's forever, never squander a chance to see what is here now. Almost two decades ago a tree house stood where only ivy now abounds. The creek has worked away at its banks, shifting, as such waters do. The white squirrel has been gone for several seasons. These changes are cyclical, and they are to be expected. Yet other changes are forever. Just seventy-five years ago, the last of the Carolina parakeets were spotted in the wild. No woods will ever again hear their call. The year of this writing marks the hundred-year anniversary of scientists identifying the fungus that would go on to ravage the once copious American chestnut, a tree that is said to have counted for one of every four in the Appalachians.

In the time that it has taken to write this, the sun has ascended into the crowns of the trees, and it has become just warm enough for a few mosquitoes to sample me. Sporadic wisps of sawdust drifting from the upraised end of a deadfall compelled me to interrupt my composition and investigate the otherwise hidden work of some carpenter ants. The crows are aflutter over something, and I am eager to see what they have turned up.

Someone with an unsound ego and a poor understanding of physics once asked, "If a tree falls in the woods and there is no one around to hear it, does it make a sound?" You bet it does. If the season's first swallowtail lights beside a puddle and there is no person around to see it, the butterfly still lights. If only raccoons taste the berries on a blackberry bush, the taste is still as sweet. If a mantis produces an egg case or a shining lizard darts across the path, or a pair of owls begins calling to one another and no person is there to observe it, the only result is that humans have missed it. Maybe that's why I keep coming back. I've realized that the question is less riddle than warning, and when the sound of a falling tree echoes through these woods, I want to be the one that's there to hear it.

A great blue heron finds security near the wetland.
~Davis

Several miles of trails in the preserve are clearly marked. *~Davis*

A towering sugarberry tree in spring raiment.
~Schmunk

Early spring affords a rare open view down the trail. *~Schmunk*

Evening sun colors the edges of "Andy's Bridge." ~*Schmunk*

Bullfrog tadpoles in the wetland. *~Schmunk*

Along the Ridge Trail broad mayapple leaves catch oak catkins and dogwood petals. *~Schmunk*

Wild roses bloom in masses on the periphery of the wetland. *~Schmunk*

Junction of the Cottonwood and Beaverdam Trails in evening light. ~Schmunk

Carolina geraniums atop the ridge. ~*Davis*

A rare columbine blooms near the end of the Cottonwood Trail. *~Schmunk*

Chairs invite visitors to linger near the pond maintained by Ed Griffin. *~Davis*

Fallen trees will enrich the soil for wildflowers on the ridge. *~Davis*

Solomon's seal graces the margins of the Ridge Trail. *~Schmunk*

Winged fruits of silver maple cover the Cottonwood Trail at the foot of the ridge. *~Davis*

An expedition of the South Carolina State Guard led by Brigadier General Ed Y. Hall prepares to clear debris from Lawson's Fork. *~Davis*

Delicate blossoms of Carolina silverbell at the foot of the ridge. ~*Davis*

Phlox blossoms cling to the bank above Lawson's Fork. *~Schmunk*

SENSES

A WALK ON THE TRAIL IS MOST GRATIFYING WHEN you commit all of your senses to it. If you really want to see something, you must look intricately. Listen deeply, too. As any birder can tell you, there's often more to be heard than seen. Moreover, due to the extraordinary directional hearing of the human ear, you might be led to see something that you would have missed otherwise.

I can never resist the impulse to stick my fingers in the soil or caress the back of a bumblebee as it gorges on nectar, and to me the difference in frogs and toads has always been in the feel. Smell, too, has its place. You can most easily identify many plants and fungi by their distinctive odor. I rarely go by a papaw without crumpling a piece of the leaf and treating myself to its rich and distinctive odor. Taste? Anyone who has stumbled upon a drooping muscadine vine, wandered into a wild blackberry thicket, or spent an afternoon engaged in the tedious art of consuming honeysuckle will likely tell you that this sense gave them the highlight of their trip.

One morning I determined that I was going to take an entire day to walk the trail. I paused with every step, watching and listening. Although my slow-motion hike amused several passersby, it was an amazing experience. At one point I watched the earth bulge at the subterranean progress of a mole, and I later witnessed a ruddy snake languorously unfurl from a mid-stream stump and pour itself into the water. The highlight of my odd journey, however, was happening upon a field mouse that was dining on seeds. So slowly did I move that I man-

Annuals planted by trail custodians attract visitors and bumblebees. *~Davis*

aged to stroke its back before it looked up, startled, and scurried into the underbrush.

As I ate lunch that day, I contemplated the senses that we do not have. How different, I wondered, would my walk have been if, like some moles I've read about, I had a keen awareness of Earth's magnetic field? Insects, of course, can see portions of the light spectrum that we cannot, and many animals hear more than the narrow range of frequencies that humans can.

Plants, too, have senses. The Cottonwood Trail's vines are quite aware of gravity, and they are capable of sensing light as well as dark. Like almost all plants, the vines need sunlight. Vines—like poison ivy, English ivy, and muscadines—are a bit more active in their search, though. They will propel themselves in a slow slither towards the shadow cast by, say, a tree. Once there, they sense the object and, in some cases, alter their growing patterns so that they can wind up whatever they have found. I have wondered if the trees in the preserve can sense the first curling embrace of a vine and, if so, whether they are capable of considering the wearying mass of their potential burden.

Vines are a conspicuous part of the canopy in much of the preserve. *~Davis*

WATER

LAWSON'S FORK STARTS AS A SPRING IN A pasture just north of Inman Mills. My family and I used to take water samples near the source, and my little girls dropped leaves into the creek's inaugural waters. "Fairy boats" they called them, and they pretended that the boats were carrying miniature guests to a royal fairy wedding.

In Lawson's Fork calligraphic patterns written by water. *~Schmunk*

Suppose that we could follow one of the fairy boats as it traveled from here to the Atlantic. After traveling Lawson's Fork for twenty miles, the diminutive crew would end up in the Pacolet River. From there they would join the Broad, and then the Congaree. If our fairy boat survived these waters, it would make its way into the Santee-Cooper, and, with a talented navigator and a lot of luck, the miniature vessel would arrive at the Atlantic via the Charleston Harbor. From start to finish their adventure would take them about two weeks.

This water has journeyed farther than even our fairy boat trip suggests. Although it would be impossible to trace the path of an individual water molecule from its inception to present, the mind boggles at the possibilities. This water is ancient, billions of years old. Thus, currently floating through this preserve are water molecules that have passed through the gut of some prehistoric beast, perhaps washed over the feet of Jesus, or trickled down the brow of Genghis Kahn. Some of this water has ventured deep into the earth, fallen on faraway lands, tumbled in an ocean on the other side of the globe. Unlike any other substance on earth, water exists naturally as a solid, liquid, or gas. Thus, some of the water that now passes before you has probably been locked in a glacier and, at another time, dispersed in the air before becoming a cloud.

Humans are only now beginning to grasp the physics of water. We still cannot wholly interpret the rhyme and reason behind its movement, and our most powerful computers are not able to predict exactly how it will behave in even relatively mundane circumstances. For good reason, too: a single drop of water contains *billions* of individual molecules! Gaze at any spot along Lawson's Fork and the number of water molecules visible to your eyes far exceeds that which the human brain is capable of comprehending. Moreover, as this water weighs about sixty-two pounds per cubic foot, you will be witness to the movement of thousands of pounds of water. I wouldn't dare to guess how much the entire creek weighs.

Water is a connection shared by all living things. Take the stately river birches that bow so gracefully over this creek—each is composed of about seventy-five percent water. That is the same percentage of water contained in your own brain. If there truly is a tie that binds, that tie is water.

Ice, mist, and creek: three forms of water. *~Davis*

DEAD TREES

THIS LATE-FEBRUARY AFTERNOON IS RAINY, AND the wind cavorts among the treetops. But it is warm. My girls, four-year-old Mae and seven-year-old Ella, and I hit the trail. It's simply not wise to squander any warmth that February offers. Nevertheless, Bonnie, my wife, stays at home with our newborn son, Finch.

We park at the baseball field, cross Sydnor Road, and walk onto the Loblolly Trail. Being beside the baseball fields would be a peculiar torture for the cognizant ash. For what is to us a wooden bat is to the ash a relation. Of course it is the nature of trees to live in the midst of their fallen comrades. Many of the pines on the Loblolly Trail are of the horizontal variety. My daughters find these to be of particular interest. They're good for clambering over, and they occasionally yield secrets.

We wander off the trail until we find a tree that suits them, a short-leaf pine broken at its base. While my daughters investigate, I measure. It's about sixty inches in circumference and one hundred feet long. The rings are wet and crumbling, but we can clearly discern seventy. It was probably eighty or so when it toppled, and, by its state of decomposition, we can judge that it's been down for a few years.

Borrowing a trick from ecologist Aldo Leopold, we can pair this old tree's rings with the human history that paralleled their formation. When this pine was a seedling, electric trolleys still traveled through downtown Spartanburg, and an incipient Converse Heights neighborhood was gaining momentum. While the county's human inhabitants were suffering through economic depression, it was thriving, and it was flourishing still when the county filled with soldiers and the country went to war.

When this tree was as old as I am at the time of this writing, neighboring land was purchased, and Spartanburg High School was built. Not much later, it probably reached the canopy. Even higher overhead flew the first airplanes to come out of the newly built Greenville-Spartanburg Airport.

This now-fallen pine stood by when in 1984 the Junior League of Spartanburg established the River Birch Trail that runs behind Spartanburg High School. Shortly afterwards, the Spartanburg County Foundation assembled a committee to address the loss of Spartanburg County's natural areas. In 1989, when this tree was a mature sixty-four years of age, the committee morphed into the land trust that is now known as SPACE (Spartanburg Area Conservancy).

In the summer of 1989, SPACE began to

consider tracts to purchase for preservation. Local biologist Jack Turner suggested the land on which this pine grew, and, after much deliberation (including observations about the enduring prevalence of mosquitoes), the group determined to purchase the property. Jack Turner and Harold Hatcher, founder of Spartanburg's public garden, directed the bulldozer that formed trails in its midst, and Dr. "Botany" Bob Powell of Converse College gave the Cottonwood Trail its name. Renowned naturalist Rudy Mancke led a group by this tree as they celebrated the trail's 1991 dedication. Then some time later, prompted perhaps by a wind, the mighty pine fell.

Under its bark the dead tree is etched with cryptic markings, the legacy of grubs. Channels of yellow hyphae course over the sapwood. The wood is soft and crumbling. Eighty years of life returning to the soil.

As we return to the trail, we step over the eroded top of the venerable tree. Here the decomposition is greater. Mae points it out first: rising from the decaying trunk are several seedlings. One day, I imagine, my children might bring their own children here. If so, perhaps they will marvel at one of these saplings, grown and fallen, nourishing yet another generation on this land. If that day should come, I hope that my children will remember to tell them the story of the Cottonwood Trail's inception and remind them that a small group of dedicated people can always make an enormous difference.

A pair of pines returning to earth along the Loblolly Trail. ~*Davis*

Intimations of Spring

Spring is more than a full month away, I know, but the daffodils in my backyard are blooming, and I cannot resist the impulse to visit the Cottonwood Trail. I'll visit twice this day—once in the morning and once in the afternoon. My children accompany me on both trips. They always notice more than I do. How I wish I retained their ability to see things with new eyes!

It's over sixty degrees. Perhaps that's why I'm not the only one that has been tricked into believing it's spring. For the first time this new year, the peepers are calling in the wetland. There will again come a time prior to spring in which the cold will silence them. Surely they know it as well as I do. Along the trail, an early gathering of bugs is posting, and I happily shoo them as I pass through their swarm. Never mind that there is more winter left, I am a willing participant in this delusion.

By the calendar of our days, we can tell that this weather is a gift. I wonder what it is that tells these woods the same. The heat? I doubt it. After all, it was freezing last night, and if morning revealed ice on my car, it was likely on the vegetation around the Cottonwood Trail, too. Nature is wiser than that. I have my own hypothesis: the angle of the sunlight. It's the fall that makes me think so. Notice, in the still-sweltering days of September, how the trees begin to drop their leaves. It's the first sign of the inevitable cold. Maybe in this time, too, six weeks before calendrical spring, the woods sense a shift in the path that our star takes across the sky. Perhaps the plants, the animals, know that there's more to come, but look to the sky for the spring that, long past the calendar date, will settle upon us.

Then again, maybe it's joy. There are those who will scoff at this, will accuse me of anthropomorphism. I invite those people to take a walk in the woods with me. I defy them to dismiss the joy—that really is the only word—apparent in the tree-to-tree capering of a pair of gray squirrels. And if this is a world in which every songbird has to sing for a purpose, you can count me out.

At the end of the day, that's what I settle on—joy. My girls run ahead of me on the trail and gather treasures such as a sprig of loblolly needles. Our old dog struggles to keep up with them, finally gives up and dines on grass. A couple of boys stand by the creek and strip off their shirts. The wetland trembles with the voices of the peepers. "Spring is coming!" they call. Joy.

A hybrid of Fowler's Toad ventures onto a timber near the Beechwood Drive parking area. *~Davis*

Crows

If you're ever knocking about the woods at a loss for what to do, I suggest listening for crows. If there's something interesting amiss, these rounders are apt to know it, and they don't mind sharing the information with anyone willing to follow their calls.

I keep a sharp ear turned for a riot of crows, and I am rarely engaged in something so pressing that I won't traipse through the brush to see what they are up to. Sometimes I arrive to find nothing save for the crows. I can't help but feel that these wily birds have turned the tables and made me the entertainment. I feel certain that crows are capable of such trickery.

Usually, however, crows find something more interesting to pick at than an unsuspecting human. The woods abound with creatures that would gladly make a banquet of a nest full of eggs or hatchlings, and a creature as intelligent and physically formidable as a crow is loathe to stand aside and let this happen. On numerous occasions, I have come upon a mob of crows worrying a black snake. I have only once seen a black snake come out the winner in this dispute. Generally, there is no retreat hasty enough when you're a snake in the midst of angry crows.

It's this scene—crows after a snake—that I thought I would see when several years ago, I sought out the source of a crow-begotten ruckus. Instead, in the midst of that black cloud, was a great-horned owl. I had never seen this species in the preserve, and I was surprised at a daylight encounter with this nocturnal raptor.

The bird was not so amused. Its initial strategy was to ignore the whole thing, two dozen shrieking crows and all. Rather than acknowledge them, it perched on a pine bough looking as disgusted as the avian countenance can effect. The only time it moved was when I produced a pitiful hoot. At this, the bird cocked his head and sized me up for a heap of owl pellets. I became hyperaware of its talons and hooted no more. After several minutes, it spread his enormous wings and soared over the treetops with the pack of crows close behind. I could hear their continued hassling long after they were out of sight.

An eastern garter snake lies in underbrush near Lawson's Fork. *~Schmunk*

TRAVELERS

AT LAST IT'S SPRING, AND EVERYTHING IS ASTIR. I recently saw a sulfur flutter past, careless, as if he'd been here all along. The woods are full of melodious proclamations reminding us, every last living creature, that this robin or that mockingbird owns tag, title, and deed to these parts (and all the females herein), and to tread lightly, for we tread upon his dominion.

A haze of color clings to the Cottonwood's treetops, wisps of smoke that rise before a conflagration. No leaves or flowers yet, but the suggestion of leaves and flowers, whispered promises of the summer sun on which they will feast, the fragrance of the air, the shade under which we will later pause.

Despite the mud and wind, this show never fails to draw a crowd. Just a month before, I could walk the entire trail and never see a soul, but I now pass a dozen people on my walk. It's a good sign, I think, that the woods remain capable of drawing a crowd. After all, between cable television, electronic game systems, the movies, and the mall, there's hefty competition. It's nice to see that nature can, at least with some folks, hold its own in the contest for our attention.

The woods, even woods as small as these, change us, and people act differently on the Cottonwood Trail. Out there—the surrounding neighborhoods, the roads, and shops—you see very little contemplation. Out there, we're too often skimming the surface of our own lives. In the woods we live more deeply. People slow down, they take notice. In the woods we scrutinize the ground and study the skies.

Late-March, and I was walking the bend on the creekside of the ridge. Stopped by the side of the trail was a man wearing a button-down shirt, khakis, tasseled loafers. He was facing the creek, and his hands were in his pockets. There he stood, eyes closed and letting the sunshine wash over his upturned face. Do that "out there," and people think you're nuts. In here, though, it's understood.

It is as if time gets caught here, gathers in the clefts of the earth, snags in the fabric of branches and underbrush. Once on a whim I began to follow an odd track. It was freshly crafted, a meandering line in the mud, the clay, the sand, and it led me for almost the entire length of the trail. Nearly two miles after I had started, I came upon a boy walking with his father. One of the boy's hands was enveloped in his father's. In the other, he held a stick, and with this trailing carelessly behind him, he was creating the line that marked our collective journey.

I appreciate a place where we can slow

down and make a record of our passage or take the time to serendipitously follow our curiosity. We need more places where we feel free to close our eyes, thrust our hands in our pockets, and remind ourselves of what it means to have our faces warmed by the sun on a springtime afternoon.

The Gauthier family enjoys an outing on the trail. *~Schmunk*

Antlions

I ONCE SPENT AN AFTERNOON TRYING TO TEASE antlions from the sand. I drew a single pine needle across their burrows and watched them toss sand at it, trying to make it falter and tumble into the grip of their upturned jaws. When I drew one of them to the surface, I dug it out and blew away the sand until only the antlion remained in my palm, backing away, tickling over my skin, trying to burrow back into the earth.

In others I could elicit no movement, and I determined that they must have been abandoned. On a whim I scooped up one of these, thinking that I might find a carcass, perhaps a molted husk, or even a particularly stalwart but healthy antlion. Instead, at the bottom of the funnel, was a cocoon. Imagine an ancient earthen jar drawn from the depths of the Mediterranean, the kind of vessel salvaged from the wrecks of wooden ships, containers that once held wine or oils for people long ago. What I found at the bottom of that burrow was exactly that perfect, that magnificently constructed. The little jar was spherical and tan and created with mathematical exactitude, and to its top was attached a hinged lid. In my hand lay a flawlessly rendered relic that, had it been made larger and by human hands, would have been a well-chosen piece for any museum.

Like any work of art, it told a story. Unnoticed by the eyes of any human, buried within the sand at the side of the Cottonwood Trail, an antlion had begun this work. For three years, this creature had made a living at the bottom of its funnel, moving only to toss sand at creatures unfortunate enough to happen into its burrow. Then, at a signal that humans will never know, it began to build. This beastly larvae, with its maggot-like torso and scimitar jaws, answered a call to do something that it had never before done, to build something that it had never before seen or been taught to create. In the obscurity of the earth, the antlion by instinct crafted one of the most elegant objects I have ever encountered. A month passed, and one evening the once-fearsome creature emerged from this tiny jar. Transformed, it took to the air on wings as intricately paneled as the glass in any cathedral and left its magnum opus behind, discarded in the sand where I found it.

In the inscrutable mind of the humblest of creatures is contained the blueprint for an object of such profound grace. I haven't the will to resist that kind of magic.

Mushrooms and English ivy on the edge of the trail. *~Schmunk*

A tiger swallowtail butterfly feasts amongst zinnias and cosmos planted in the clearing beneath the power lines. *~Davis*

Hunt Russell, a Sunday afternoon fisherman, holds a largemouth bass from Lawson's Fork.
~Davis

Cosmos blossoms attract tiger swallowtail butterflies. *~Schmunk*

Irises at the end of the wetland boardwalk. *~Schmunk*

Lady ferns in the shade near a feeder creek running south of the power lines. ~ *Davis*

Winged fruits shrivel on a silver maple uprooted by a flood. ~*Davis*

Zinnias and cosmos blossom below the Duke Power lines.
~Schmunk

Arrowhead leaves surround a fallen log in the wetland. *~Schmunk*

Dense summer foliage surrounds a bird box in the wetland. *~Schmunk*

Cardinal flower on the edge of the wetland.
~Schmunk

Clematis in blossom and cinnamon vine: two invasive plants along the trail. ~Schmunk

Trail custodians invite visitors to feed the fish from the Duke Power Bridge over Lawson's Fork. *~Davis*

Puddles on the muddy trail in deep summer. *~Schmunk*

A cluster of fungi at the base of a creek-side tree. *~Davis*

Cinnamon vine, an invasive species, thrives near the Loblolly Trail. *~Davis*

A summer drought allows a view of the Duke Power Bridge from a downstream sandbar. *~Davis*

FLOOD

SEEING LAWSON'S FORK AT FLOOD STAGE IS astonishing. In the last twenty years I have seen numerous floods. In my experience the most affecting sight is when the creek is on the cusp of spilling over its banks. At that point, this unassuming waterway becomes a new beast, one that devours entire sections of creek bed, toppling trees and carelessly tossing logs and debris downstream.

One imagines that this creek, or any flowing body of water, at flood stage would make a lot of noise, the rumble of whitewater. Not so. The creek at flood stage sounds exactly like the creek at any other time. There's something about the lilting tinkle of water that seems more ominous when the creek is feeling its oats. It's almost subversive.

I love the language of a June 6, 1903 report in *The Spartanburg Journal* describing a flood of Lawson's Fork as "probably the most serious calamity in (the Spartanburg area's) history. The rivers overflowed their banks to heights never before known, the creeks became big rivers, and the small branches and rivulets became roaring torrents." After telling how Lawson's Fork demolished the Southern Railway trestle that had once straddled it, the exuberant author claims that the waters of the creek "could be seen rushing madly downstream almost like a thing of life."

About ten years ago I ventured into these woods after a few days of particularly heavy rain. In some areas water blanketed the trail, but the higher ground was relatively dry, and I was able to make my way down to the creek. Lawson's Fork was level with the bank and churning by at a frightening clip. Everything about it spoke to the enormous weight of the water and the momentum of such immense volume.

I was standing right beside the towering cottonwood tree that grows beside the Duke

A victim of flooding straddles Lawson's Fork. *~Schmunk*

Power Bridge when the creek capitalized on the opportunity to run amuck. You can still see where a section of the bank gave way, though it's recently been filled in with bricks and concrete. It seemed to me that it made a noise like a cannon as an enormous amount of earth burst into the air and a torrent of water came rushing behind it. Alarmed, I made a hasty retreat down the Pinecone Trail. The creek beat me, though, and I didn't get out of the woods without being up to my ankles in water.

You might wonder where the resident animals go in the period before the creek recedes, leaving mud and tons of disgorged sand in its wake. After a flood in the 1980s two of my friends and I headed to the preserve with our rafts. We had floated well into the woods before one of my friends pondered a version of that question aloud, "Where do you guess the snakes go when it gets like this?" As if prompted by the query, a thick brown snake appeared and headed our way. I do not know what its intentions were, but if it had in mind to share one of our rafts, none of us cared to oblige. I paddled so fast that my arms ached for two days.

Receding floodwater leaves a bluegill stranded on the trail. *~Schmunk*

Pinecone Trail

I have always had a soft spot for turtles. When I was a kid, I would frequently bring them home. My mother was a good sport about this, but she established some parameters. She allowed me to keep the turtles for a day; afterwards, I had to release each turtle at the spot where I had found it. Since I became devoted to every living creature that I caught, this policy broke my heart. That's where my mother's true genius came to bear. She explained that the turtles both needed and wanted to live in the woods, but, as a concession, I could use her fingernail polish to paint my initials on the belly of "my" turtles.

An eastern box turtle hides along the edge of Lawson's Fork. *~Schmunk*

I was surprised at how often and for how long I continued to come across turtles that I had tagged. Knowing a little more about our local turtles, though, explains a lot. The Cottonwood's native turtles tend to stay in a narrow range, and some, like the eastern box turtle, can live for a full century.

The Pinecone Trail is a good turtle-watching spot. Both the eastern painted turtle and the eastern box turtle like wet conditions, and the nearby wetland offers suitable habitat for snapping turtles. The snappers tend to hang out with everything but their faces submerged in the water, though they do sometimes venture onto land. While on an early morning walk around the wetland, some friends of mine encountered one cantankerous snapper that refused to give ground. Rather than tangle with him, they gave him wide berth. As snappers have strong jaws, a deceptively quick lunge, and seem to be in a perpetual ill humor, this was a wise decision.

Turtles aside, the Pinecone Trail is one of my favorite parts of the preserve. It's not as well-traveled as the other trails, so it's a fine place for solitudinous reflection. Since the earth here is usually wet, it holds tracks well. Walking on the trail today, I saw several raccoon prints, some bird tracks, and side-by-side prints of an adult whitetail deer and its fawn. A recent storm has made a mess of the trail. As I slipped along, I noticed that some of the deer prints, too, were skewed, suggesting that they didn't have much better luck with their footing than I. No wonder—the loblollies bear water marks that testify to the depth of the recent flood. I stood by one hale conifer and measured the ochre-colored line with a modified salute. It was right at nose level, so, at its highest, the water here was over five feet deep.

My goal on this day was to relocate an eastern painted turtle that my wife and children rescued from a busy road. He was eager to be released, but I had one piece of business with him before I let him go. Grasping his carapace and trying to avoid his formidable claws, I turned him over. If you're ever walking along the Pinecone Trail and run into this turtle, you'll know he was the one. On his plastron, painted in hot pink, are my initials.

Privet rises among fallen leaves and cones, along the Loblolly Trail. *~Schmunk*

The Battle

The root of every war is resources. Out here, most of the fuss is over one resource: energy. Not just any energy, but starlight from ninety-six million miles away, rushing through the cold emptiness of space at 186,300 miles per second. Be you plant or animal, fungi or protist, this starlight is what it's all about.

On some dry, warm day, go to the Cottonwood Trail, lie on your back, and stare up at the sky. Or try to, at least. You'll be lucky if you find a patch of sky that a leaf hasn't already taken. There's heavy competition for the placement of these solar cells. Some trees, like the tulip poplar, use the weapon of height, rocketing into the upper reaches of the canopy where the light is abundant. Others, like the river birch, use their powerful roots to entrench themselves along the banks of the creek. From this position, they lean out over the water where no other tree shades them. Vines sap the strength of their opponents, and, where a tree has fallen, the ground suddenly springs to life with competing undergrowth.

One afternoon I was hanging out by Beaver Dam Trail, admiring the handiwork of the industrious rodents for whom the trail is named. From this position I was privy to an impressive display of aerial combat. From the canopy above me came a branch-shattering crash. Looking up, I saw a sparrow headed right towards me. It was not alone; close on its tail was a hawk. The sparrow was in a full dive, and the hawk had its enormous wings extended by its sides, talons outstretched. It seemed a calculated move by the sparrow, and, if so, it took a lot of guts, for it didn't pull out of its dive until it was about two feet from my upturned face. The hawk lost this game of chicken, and retreated—so close that I could feel the air from its wings—as the sparrow escaped to the underbrush.

I have seen rabbits and squirrels that have had less fortunate encounters with hawks. I have seen a great blue heron here score a fish, and a snake stalk a sunning frog. There are trees in the preserve that sport opportunistic fungi and others that are choking under the weight of vines. Too, there are trees that bear evidence of winning their slow battles, trunks bearing enormous twisted scars. It's a war out here, and it's easy to want to take sides. I'll admit, for instance, that I was pulling for the sparrow. At the same time, I would hate to think of these woods without hawks. As my wife and I tell our children, everyone has to make a living, and out here, quiet though it may seem, that living is by tooth, thorn, and talon.

Tulip poplar and pawpaw trees flourish in their forest niches.
~Schmunk

Bridge Crossing

Although the entire trail is in proximity to "development," this place brings the trail most obviously in contact with the manmade world. At this spot where Fernwood-Glendale Road crosses over the trail, man and nature cross paths. This is the only place on the trail that is devoid of vegetation. From the beginning of the bridge to its end, hard packed clay is the predominant feature. Lack of sunlight is one factor. Another is the bombardment of runoff that falls from the bridge. When it hits this earth, there are no plants to break its fall. Not even a riparian zone (a row of vegetation along the creek) has been left intact. The runoff has eroded the soil and in some places carved deep channels in the clay. All of this runoff, and all of the accumulated exhaust, trash, and soil that it has picked up along the way, ends up in Lawson's Fork Creek.

April 11, 2002, was a SPACE cleanup day. On that rainy morning a couple of dozen volunteers waded into the creek to retrieve the trash that burdens this waterway. Most of the group started at Spartanburg High School and worked their way downstream, but I chose to start at the section of water underneath this bridge. In just two hours of cleaning a swath of creek about one hundred fifty yards long, I retrieved one glove, one milk crate, one plastic chair, one baseball bat, three car tires (one tire was so well lodged I couldn't get it), one bike tire, one partially full paint can, six plastic soda bottles, one plastic oil container, one milk jug, two plastic balls, one shoe, one shotgun shell, one shoe sole, one glass bottle, one toilet seat, one plastic raft, one inner tube, one pizza box, one plastic bleach bottle, one battery, one calculator, myriad plastic bags, Styrofoam pieces, broken bottles, odd containers (chip bag, syrup packet, etc.), and seventy-three beer/soda cans (some aluminum, some steel).

I pulled one beer can from my heap and mailed it off to its manufacturer. Soon after, a representative of that company called me and told me that they manufactured that can between 1971 and 1973. In other words, thirty years prior, some thoughtless soul celebrated the completion of a cold one by chucking the can somewhere that it didn't belong. Maybe that person, probably now in his or her fifties, tossed it in the creek. Maybe that person threw it on some road miles from where I found it, and it washed to these waters via stormwater drains. Either way, it reinforces ecologist Ernest Callenbach's statement that "everything goes someplace." Trash doesn't disappear just because we throw it out.

Thanks to tighter environmental regulations and groups like SPACE, Lawson's Fork is healthier than it has been in a long time. When I was a kid, I never saw fish in these waters, but I'm seeing them now. Dragonflies, a positive environmental indicator, also seem to be on the upswing. It's a step in the right direction, but we have a ways yet to go.

Evidence of human impact: a doll's head surfaced on a sandbar. ~*Davis*

THE CHALLENGE

THE SOUND OF A BLARING KLAXON HAS BARELY left my ears, but already there is more clamor: “Drop and give me ten push-ups! Team Captain counts them!” Claire, Curt, Marcel, and I, four thirty-something friends who are feeling older by the minute, drop into push-up position, and I begin to shout off the count.

It’s a late-April morning on the Cottonwood, and our team, Mama and the Papas, is competing in the Second Annual MaxMotivation Upstate Ultimate Challenge Run. Former Marine Derrick Singleton manages the event. He takes the Marine part seriously, making sure it’s no ordinary 5K—plenty of yelling, water, mud, and as many forms of exertion as he can pack into the prescribed distance.

The trail frequently hosts those who are seeking to challenge themselves. It’s one of Spartanburg’s most beloved places to run, walk, and bike. Mary Walter, SPACE’s executive director, thinks she knows why. “The Spartanburg Running Club members and other runners really appreciate our trails because it gives them a break from running on pavement. Much better for the legs!” Besides, she says, the trail offers something that the surrounding roads may not. “Every time I go to the preserve, I discover something new. All it takes is curiosity and interest.”

Having succeeded in carrying Claire up a set of stairs, we approach another station. Marcel sprints away to accomplish the task. Soon, we’re back in the water, carrying weights across the creek.

The mission of SPACE is “to protect and preserve natural areas of ecological, historical, and aesthetic value to enhance the quality of life for all residents and future generations.” As part of addressing this mission, SPACE holds an annual 5K called the SPACE Race. This event usually draws about sixty runners, and in 2004 a family of five completed the race together.

We’re almost to the ridge when we realize that we’re making competitive time. We use a rope to pull ourselves up the backside of the ridge and then race around the front of it. Marcel and Curt carry Claire and me over thirty yards of sand, and then we try to keep up with Marcel as he cruises over the Pinecone Trail. Stopped again, we repeatedly lift a tractor tire.

By the time we finish, all of us are soaked and mud-streaked. We wash up while we wait for the remaining teams to complete the course. Everyone gathers for the results, and we learn that with our time of 41:50 we’ve earned third place.

After we’ve admired our trophy and promissed one another that we'll take first place the next year, I ask Derrick why he uses the Cottonwood as the site of the Ultimate Challenge Run. "It offers a tremendous amount of natural obstacles," he says. Furthermore, like any good Marine, he knows a solid mission when he sees one. "I believe with continued awareness of this great treasure in Spartanburg it will greatly benefit the community, local businesses, and enhance the lifestyles for people from the entire upstate." Well said, soldier.

A solitary walker heads toward the far end of the trail. ~*Schmunk*

A Catalog of Spring

On a Sunday afternoon in May, Bonnie and I take the children to the trail. We carry Finch in our arms and follow our daughters, my wife collecting the shoes that they kick aside. Trumpet vine is blooming on the Loblolly Trail, and, the week before, I witnessed the play of a pair of whitetail deer on that side of the Cottonwood. I'd like to head that way, kick around amongst the fallen tulip poplar flowers, but my girls are eager to see what they can turn up near the wetland.

Christmas fern fiddleheads on a hillside below the Ridge Trail. *~Schmunk*

Preparations for summer are well under way. The bloodroot has come and gone along the ridge, and mayapples will soon begin to bloom. This is usually the month in which I note the first tentative flashing of lightning bugs. The underbrush has thickened again, and from its depths rise the white-petaled anemone of blackberry blossoms. During the previous week I startled a cottontail that was dining atop the ridge, and there are rumors of columbine further along the trail.

Closer to the wetland, it's a veritable blizzard of cottonwood seeds. Tadpoles speckle the shallows. My daughters run over the Duke Power Bridge and sit in the sand. "Will it snow again before summer?" Mae asks as cotton falls around us. Bonnie shakes her head no. My daughter shrugs, lies back in the sand, and slowly waves her arms and legs back and forth to make a "sand angel."

At some point in our lives, somewhere along the route to adulthood, we learn to dismiss the beckoning whisper of flowing water, that chiming cadence that invites us to wade in, let the surface pass through our trailing fingers. But what our ears fail to hear is a clamor in the heads of my girls. Seated in the sand, the children ask if they can wade in the creek. "The air is warm, but the water is too cold," we insist. The girls stand wordlessly gazing by the water's edge. Cottonwood fluff dimples the surface, and the shadows of bubbles and invisible currents spread like smoke over the creek bed. Who are we to separate our children from this breed of wonder? "Okay," we tell them, "but only up to your ankles." Immediately, they're in up to their calves, and their laughter skips over the water like thrown stones.

Through the gentle heat comes the lazy buzz of insects and the hollow call of frogs. My oldest daughter pulls a bottle from the creek.

What is this thing that we call spring? More than a season, beyond the tilt of the earth or a place in our calendars. Spring is the barefoot tracks of children and a single dark snake sunning on a stump. It's the girth of the tadpoles, the thickening of stalks. Spring is the beckoning of flowing waters, and it's the mass of sunlight added to the air. It is the one time of year when, for a few fleeting weeks, anything seems possible.

"Want to help?" our daughters ask, prying two more bottles from the sand. Bonnie and I place our shoes on the bank and step into the water. It *is* too cold, but it is wonderful as well. "Are we saving the planet, Mom?" Ella asks, discarded bottles in hand, her hair alive with sunlight.

"Yes," says my wife. And in the tingling of the creek and air infused with the silvery wink of drifting cotton, we all believe it.

Bloodroots herald springtime on the south slope of the ridge. ~*Davis*

An arch formed by a tree over the Cottonwood Trail near the ridge. *~Davis*

Swirling water of Lawson's Fork blurs the reflection of fall foliage. *~Davis*

Fall color from the southern end of the Duke Power Bridge. *~Davis*

Ripples and reflections along the edge of Lawson's Fork. *~Schmunk*

Black walnut fruits and leaves lie on the Cottonwood Trail. *~Schmunk*

Ed Griffin's tractor treads with leaves on the Turkey Cut Trail. *~Davis*

A pearl crescent butterfly enjoys the nectar of goldenrod. *~Davis*

At the Fernwood Drive bridge crossing of the Cottonwood Trail and Lawson's Fork. *~Schmunk*

A basswood leaf and mushrooms, temporary companions on the trail. *~Schmurk*

A daddy longlegs blends with the bark of a tree. *~Davis*

The serrations of cottonwood leaves cast shadows on a fall afternoon. *~Davis*

A tiny beech involucre balances on a pawpaw leaf. *~Schmunk*

The feather of a yellow-shafted flicker brightens the wetland groundcover.
~Schmunk

A great blue heron seeks a meal. *~Davis*

As Ed Griffin watches, Lynn Rhodes removes a fallen tree from the Loblolly Trail. *~Schmunk*

A hint of fall color appears on foliage at the Ridge Trail junction. *~Schmunk*

Cattails surround the open center of the wetland. *~Schmunk*

Freshly fallen red maple leaves exhibit vivid colors. *~Davis*

Wetland

Getting here used to take commitment. There was a time when the wetland was accessed only via a tightrope walk of fortuitously assembled deadfalls. It was a route fraught with adventure, keeping your balance while simultaneously dodging raccoon scat and blackberry thorn. But the terminus of this log trail was where the real commitment began—it was either the end of the line or a sometimes waist-deep slog to the interior.

Even a veteran muck-wader had to welcome the 2004 addition of a four hundred and thirty-foot boardwalk. Same payoff, less mud. Aside from the fact that more people can now see this marvel, the clothes-washing spouses of those who insist on seeing it, boardwalk or not, are happier.

As evidenced by the state of the trees in the wetland, it has been here for quite a while. While the rest of the preserve is flooded only periodically, the wetland is consistently covered with water. Some wetlands aren't comprised of standing water but are made instead of constantly wet soil. At times, this one has little surface water, but I've never seen it utterly dry.

Two factors contributed to the making of this wetland. First, a creek or two feeds the area. It is difficult to say just how many creeks flow into it because, in some places, they run underground. The other contributing factor is beavers. For some reason, perhaps they were trapped out or maybe they moved on their own volition (I suspect the former), the beavers have gone. Over the last couple of years their dams fell into disrepair and, for the most part, washed away. What persists is a spectacular offering: several acres of wetland that hosts flora and fauna not found anywhere else in the preserve.

On a summer afternoon I enter the wetland on the ridge-side. With each step, my borrowed waders exhale hot air, and the eight or so inches of water is quite warm to the touch. Several shrub-like plants—mallow, alder, burrweed, and black willow—distinguish this side of the wetland. This time of year the mallow bears white hibiscus-like blooms. In this setting they seem as unlikely and as bright as bone china at a picnic. A water snake pauses to regard at me before drifting soundlessly away.

Like all wetlands, these work to filter the water that comes into them. For that reason alone, they are an enormous asset. My feet sink into almost a foot of fine sediment before finding purchase, and as I approach the boardwalk, dragonflies, sometimes in mating pairs, flit by and drifts of minnows move through the miniature channels. Closer to the boardwalk, new flora and fauna dominate. Amongst the cattails I find a bullfrog pollywog that's as big as a walnut. Once in this section I happened upon several green frogs clinging to the grasses like unripe fruit, and, earlier in the summer, the *Sagittaria* bore fragile three-petaled white blooms. Here and there, the water is still dotted with the purple bloom of pickerel weed.

On the Beechwood Drive side of the boardwalk there are more stands of grass and standing dead trees. Here the water tends to move in deeper channels, and well-worn animal trails lace through the higher areas. Out of sight of the boardwalk I startle a pair of ducks.

Though the shadows have lengthened, the heat retains its vigor. As I walk back to the boardwalk, I contemplate the utter beauty of my surroundings and compare it to the retention ponds that are used to mitigate the consequences of development that too frequently abolish natural wetlands. In northern Spartanburg County, I saw a particularly large retention pond to which cattails had been added. Otherwise, it bore no evidence of life. I am reminded of a Luna Leopold quote: "The real need in conservation is the protection of those commodities like solitude and sunsets which your adding machine may tell you are economically valueless but which your heart tells you are beyond price." That sounds about right, I think, as I pull off my waders. Many of us, I think, have found in our hearts that this rare place is beyond value.

Arrowleaf and pickerelweed in the wetland. *~Schmunk*

HOLES

AS I WALK UP A TRIBUTARY CREEK OF LAWSON'S Fork, a water skimmer glides just ahead of me. To the south, a late-afternoon thunderstorm rumbles, and the cicadas sing their summer love song. A damselfly has taken a peculiar interest in my progress, and she darts from perch to perch to see what I am up to. Her classic-Camaro paint job winks through the humidity.

Few things, I think, are more beautiful than a sandy creek bed. The water has swept the sand into an impressive array of dunes. They are a miniaturized, submerged Sahara. From my perspective they look like the sort of pictures that NASA sends back from Mars: evidence of water. Large chunks of the surrounding bank have fallen off, leaving straight-edged earth in their absence. Here, in this light, they are striking, stark like an Ansel Adams portrait.

Animals and humans leave their marks in the sand. ~*Davis*

I've come here to investigate holes. Cavities are an important part of this ecosystem, and without them many of the local inhabitants would lack a home. Holes mottle the bank. Some are no wider than my pinky, while others would accommodate a melon. I probe one of the smaller ones with my finger, think better of it and continue with a stick. After all, a couple of types of wasps make their living out of holes. This one curves and extends beyond both sight and the range of my instrument. No telling how deep it is or what if anything, resides in its depths.

Sometimes I find toads perched in these niches. Perhaps they dig them out themselves, but, because of the red clay, I'd be surprised. I have seen an eastern painted turtle burrow into the soil, and the belted kingfisher digs out cavities in which it nests. Perhaps the toads capitalize on the work of others.

Farther up, raccoon tracks adorn the water's edge. I can see why: here there are shallow pools in which minnows and tadpoles swim, a regular raccoon buffet. The tracks converge on a deep hollow at the base of a tree. Another hole. No burrowing here; these raccoons have squatters' rights to found property.

There are holes up higher, I'm sure. Woodpeckers often carve out hollows in dead or dying trees. When the woodpeckers aren't utilizing them, other creatures are.

Retracing my steps, I emerge onto the banks of Lawson's Fork. There are some truly big holes here. It is possible that some of the area's aquatic mammals are the architects of these. Muskrats live in the preserve, and I have sometimes seen river otters swimming near here. I peer into one exceptionally large hole. There are no tracks at the entrance, no scratch marks. I direct my flashlight into its depths, but no shining eyes peer back at me.

A crack of thunder causes me to abandon the mystery. I scramble back to my car and drive back to my house, leaving behind innumerable holes that are, no doubt, keeping their inhabitants as dry as my own shelter.

Flying squirrels share a hole in a dead tree visible from the wetland boardwalk. ~*Davis*

FLUX

CHANGE IS THE MOST FUNDAMENTAL ASPECT of a floodplain. The primary architect of this change is water. Everything about the preserve's lowlands, from the contours of the land to what grows and lives on it, is dictated by the fact that water periodically overcomes it.

It is difficult to imagine the immense energy of moving water. Consider, for instance, that less than a foot of rapidly moving water is enough to sweep a person from his feet, and just a couple of feet of relatively slow-moving water is enough to float an SUV.

It is ironic that all of this change has borne a significant constant. Because of this flux, the tenuous balance between destruction and regeneration, the population of a floodplain is necessarily a hardy lot. And because only certain plants and animals fit the bill, they are likely to be the unflagging residents. That may seem evident, but it's not. Most forests go through a gradual transformation called succession. Over time the plants change, and as a result the animals that the forest can support change too. But not here. Cottonwoods, beeches, hackberries, maples and ashes are particularly well suited to these bottomland conditions, and they will likely remain.

Consider the differences between the flora and fauna along Lawson's Fork and that of the ridge. The lowlands are dense and populated with trees that are both fast-growing and able to establish themselves in wet conditions. The ridge, on the other hand, is less densely populated with underbrush. Oaks and mayapples thrive on this bastion from the perennial floods. You might notice, too, that the thick canopy of the ridge provides more shade and, as a result, the area hosts fewer of the non-native species that plague much of the lowlands.

Sometimes I marvel that anything survives in the floodplain. Several times I have walked the Cottonwood after particularly impressive floods and imagined that the area would be unable to rally. A few years back, flood-tossed sand obscured long portions of the trail. One of the bridges was pulled from its foundations and cast aside as carelessly as the litter that lay around it, and the nearby bank was so diminished that a segment of the trail was no longer navigable. Felled were trees with roots reaching to my deepest memories of the trail. Prospects for renewal seemed dim.

"With nature," wrote Archibald Rutledge, "ruin is transient." That lesson resides at the heart of this place. Some of what nature tore asunder, man rebuilt. The trail was rerouted, the bridge repaired. The rest was in its own way healed too. Where the bank had given way and where the trees had been defeated, plants again began to grow. Lawson's Fork followed its revised path as easily as it had followed its former course. The changes that so unsettled us were merely routine for these woods.

A pearl crescent butterfly lies dead among pebbles. ~*Davis*

ODDITIES

I GREW UP IN A UNIVERSE THAT, AS FAR AS astronomers knew, harbored only nine planets. It seems that Creation has grown a bit more crowded, even if there is room to spare. I recently read a headline stating that astronomers are finding extra-solar planets at such a quick pace that they are having difficulty keeping up with them.

Every extra-solar planet increases the probability that there is extra-solar life. Like most kids, I devoted an inordinate amount of time to imagining what alien life would look like. I have wondered, staring into the night, if there is a sentient being peering back in our direction. Perhaps that creature, too, speculates about what bizarre forms of life our planet might hold. If so, I do not believe that we will disappoint.

The Cottonwood Trail harbors a slew of oddities. My favorites by far are the fungi. What's not to like about the bearded tooth, a shaggy giant of a fungus that hangs like a disembodied beard from a deciduous tree? On the other hand, the columned stinkhorn that I once found on the trailside offered plenty by which to be offended. Unlike many mushrooms, this fungus doesn't count on the air to disperse its spores. Instead, by stinking of rotting flesh, it compels flies to land on it. Thus, the legs of flies spread its spores.

Speaking of flies, insects hold their own in the weird department. The tiger swallowtail lays its eggs on the preserve's tulip poplars. These eggs are like miniature green pearls, and from each emerges a caterpillar that eats both cottonwood and tulip poplar. Forget the standard notion of a caterpillar, though; these larvae emerge looking like a dollop of bird excrement. As bird excrement has few natural predators, it's a crafty ploy.

Another favorite is the firefly. Imagine attracting a date via a flashlight tied to your

A turtle burrows in the orange water of a stream along the eastern border of the wetland. *~Davis*

posterior. The ephemeral adult stage of the lightning bug does exactly this. The larval stage of these insects resides for years in the soil. Then, for a period that lasts just a few weeks, the adults take to the air and engage in a coruscating single's hotline. Says *Smithsonian* magazine, "They spend their adult lives courting and mating—they don't even stop to eat."

On the forest floor, keeping its compound eyes on the lambent love fest of the firefly, a nocturnal predator stalks. The Carolina wolf spider often eschews the standard wait-and-see attitude of most of their eight-legged counterparts and goes on the prowl. Neither the beating of wings nor the vibration of footfall fails to pass its notice. Several times, I have been walking the Cottonwood at night and seen their eyes aflame in my flashlight beam. It always strikes me as somewhat chilling to be sized up by so formidable a predator.

If there is an alien creature contemplating the nature of life on Earth, I would be surprised if it imagined anything as odd as what we take for granted on the Cottonwood.

Mist swirls from a wetland trunk the morning after an ice storm. *~Davis*

Invasive Plants

If you've been on the trail, you know them—privet, English ivy, eleagnus. A prominent feature of the preserve is the proliferation of invasive species, non-native flora or fauna that can imperil the health of the other components of an ecosystem. On the Cottonwood Trail the invasives tend to be plants. Dr. Gillian Newberry, professor of biology and curator of the herbarium at USC Upstate calls these "the cockroaches of the plant world—those that can coexist with man and prosper on disturbance." Disturbance abounds in a floodplain. Thus, eleagnus prospers at the trail's edge, privet stifles new growth in the interior, and Tree-of-Heaven stands at the ready to unleash an onslaught of winged seeds. By Dr. Newberry's most recent count, twenty-four percent of the plants along the Cottonwood Trail are non-native species.

Most of the Cottonwood's non-native species come from either Asia or Europe. Says Dr. Newberry, "This is because those parts of the world and ours share a similar biogeography." That is to say that our weather and growing conditions closely reflect those of the plants' native homelands. Therefore, when these plants come to our area as, for instance, ornamentals, they find it to their liking. On the Cottonwood the non-native grasses tend to be European, and the non-native shrubs and vines tend to be Asian.

It's easy to glance at the beautifully rendered leaves of *Microstegium* as it multiplies along the trailside and shrug your shoulders at the threat of non-native species. After all, a plant is a plant, right? Right—if all you're looking for is something green. If you care, however, about continuing to see old friends, plant and animal alike, then it may be time to reconsider.

The thing about non-natives is that there is little to stop them. The local plants haven't developed the ability to cope with them, and they often lack predators that might otherwise limit their success. This gives them an unusual advantage because they have no natural checks on their growth. And grow they do. Look at any of the area's kudzu infested roadsides, and there's little question about who is winning the war.

If we fail to address these invaders, the preserve will have a decreasing diversity of flora. Worse yet, an area's flora dictates its fauna. To a number, the resident animals are here because the preserve contains the plants that support them. Get rid of the native plants, and there is a good chance that you have also gotten rid of the native animals. Sure, something will fill the void; that's the way nature works. However, what fills the void will no longer be those plants and animals that we think of as making our area unique

Efforts are being made to address the threat. A Wofford College project has cleared invasives from some areas along the Cottonwood Trail to see what natives might be able to take hold. SPACE has helped to organize teams to combat the invasive species. Volunteers carry out the majority of Cottonwood maintenance, and it will take a monumental effort to defend the preserve. If you care to help, contact SPACE and meet these invaders head-on at the frontlines.

English ivy clings to a pine tree along the Pinecone Trail. *~Davis*

Snow

Autumn has passed, and with it have gone the shining yellows of ash, birch, and hickory, the sunset reds of the maples and oaks. The strength of the days has waned, and that wonderful cool, wet time when the southern outdoors smells as clean and rich as puppy's breath has gone. The preserve's hardwoods huddle together empty-handed. Some of them lift beseeching fingers towards a pale sky. Others, stripped of their warm-weather finery, magnify their dignity, easy and natural for having revealed the scars that summer hides.

We wake this morning to snow and ice. My family and I spend the day sledding, visiting friends over good winter food and cocoa. As we leave our friends' house, amidst the cellophane tapping of sleet falling onto the snow, my friend Marcel calls from his steps. "Once the kids are in bed, you want to go to the trail?"

After the excitement the kids fall asleep late. It's nearly 10 p.m. when Marcel's ancient Honda crawls up my street. "The Subaru has four-wheel drive," he says as I open the door, "but it'd be a shame to wreck it. This thing—who cares?"

We make it to the ball field, slide to a stop, and step into the night. Our breaths exit like pirouetting specters and dance into the darkness. We step into the woods, pass under the bridge, and walk onto the trail. The air is too cold to carry smells, and the scent is as monochrome as the surrounding landscape. Sleet clatters in the branches above us.

The trail is a uniform swath of snow. Our steps become indecipherable scribbles over a shining tableau. Ice-encrusted branches bow over the trail, and we walk through them like honored guests at some exotic ritual. They hiss and tremble at our passing.

By this time of year the area's reptiles and amphibians have found shelter—buried in earthen burrows or settled into the mud. In this weather it appears that the mammals, too, have taken cover. I find no tracks, but the fresh snow could be obscuring the prints of diurnal animals. Our area doesn't get cold enough for them to truly hibernate, though many have slowed down to a barely active state. Above us, unseen, mixed flocks of birds huddle in the branches, seeking warmth amongst next season's enemies.

We enter the field that borders the wetland and stop. Everything here is silent; even the sleet has paused to marvel at it. The air is imbued with unusual phosphorescence, lacking the strength of sunlight or even sunlight's weak cousin, moonlight.

This light draws no color from that over which it passes, yet it seems to illumine fully. The trees, the creek, our faces, all of them are variations on silver. It is beautiful.

We pause again at the creek and gather snowballs. These we launch into the creek, watching them bob along until they disappear. Are they consumed, we wonder, by warmth or by distance? We cannot tell. We take to the Duke Power Bridge, throwing the snowballs at any uttered target, knocking snow from branches and deadfalls.

The sleet starts again. It is past midnight, but even at this hour we are reluctant to depart. Silence and night-borne radiance are too extraordinary to leave without regret. With a final, silent pause to admire this ephemeral condition, we turn again for home, knowing that the unique conditions of this offering mean that we will not likely pass this way again.

Designs created by thin sheets of ice in the wetland. *~Davis*

Morning mist over the icy wetland. *~Davis*

An icy crust covers the wetland in the aftermath of a winter storm.
~Schmunk

Mushroom gills. *~Davis*

Inclement weather provides special opportunities. *~Davis*

Puddles persist on connector trail from Pinecone Trail to Cottonwood Trail. *~Davis*

Nandina berries provide rare winter color on the south slope of the ridge. *~Davis*

Natural and human debris at the edge of Lawson's Fork. *~Schmunk*

Bird box and cattails in the wetland. *~Schmunk*

Footbridge under the Duke Power lines in December sunshine.
~Schmunk

Trees and vines intertwine at the junction of the Cottonwood and Loblolly Trails. *~Schmunk*

Mud-stained trunks on the western edge of the wetland mark the extent of a recent flood. ~*Davis*

Fallen trees and eroded banks: the effects of heavy rain in a floodplain. *~Schmunk*

View down Lawson's Fork from the Fernwood Drive bridge *~Schmunk*

December ice melts on
a bench at the end of Ridge Trail.
~Schmunk

Electrical towers in the wetland during an ice storm. *~Davis*

Frost-edged cottonwood leaves in the wetland.
~Schmunk

In a broader sense, the West Nile virus will continue to alter the bird population. Nevertheless, with the addition of new duck boxes, the preserve's continued function as a migratory corridor, and habitat destruction elsewhere, the preserve's bird population is likely to increase and diversify.

As asphalt covers more of Spartanburg County, runoff will increase. Maxcy Lynn has already observed this trend. He is a resident of Greenbriar Road, a street that borders the preserve. Lynn has been watching rainfall amounts and the flood frequency of Lawson's Fork Creek for some time. "What we're seeing is with less rainfall, the water comes up higher and is slower to recede," he reports. "I think it's from the development upstream." To curb this trend, the county will need to follow the city's lead and establish a law requiring riparian buffer zones. This would leave creekside vegetation intact and would serve the dual function of improving water quality and diminishing run-off. In addition, the area must stringently enforce current laws to curb run-off.

As a result of natural processes, the creek will continue to shift. With it the trails will also move. Ed Griffin, for whom the preserve is named, is in the process of establishing a Frisbee golf course on parts

A sign of human presence on a beech tree at the end of the trail. *~Schmunk*

of the trail. He believes that doing so will attract more young people to the trail, young people who will then develop an interest in the trail that will safeguard its future prosperity.

I like to imagine a day when the beavers return to Beaver Dam Creek and a day when children from local schools are using the Cottonwood as the classroom that it is. If SPACE realizes its efforts, there may one day be a contiguous piece of land that spans the distance from the Edward M. Griffin Preserve to Glendale Shoals. Imagine Spartanburg preserving a swath of land that large—what a triumph! Regardless, the Cottonwood will remain a testament to the ability of committed citizens to make a difference. It will continue to serve as indisputable proof that our lives and our community are enriched by preserving the land.

Distance meter in hand, Fred Parrish checks the status of water in the wetland. *~Schmunk*

Trail Plants

Compiled by Gillian Newberry

Herbs

COMMON NAME	SCIENTIFIC NAME
Angelica, mountain	Angelica triquinata
Arrowhead	Sagittaria latifolia
Aster, frost	Symphyotrichium pilosum (A. pilosus)
Aster, small white	Symphyotrichium racemosum (A. vimineus)
Aster, swamp	Symphyotrichium puniceum (Aster)
Avens	Geum canadense
Bearfoot	Smallanthus uvedalia (Polymnia)
Bedstraw, southern forest	Galium circaezans
Bedstraw, southern three-lobed	Galium tinctorum
Beefsteak plant, common	Perilla frutescens var. frutescens
Bittercress, hairy	Cardamine hirsuta
Bloodroot	Sanguinaria canadensis
Bugleweed,Virginia	Lycopus virginicus
Buttercup, kidneyleaf	Ranunculus abortivus
Buttonweed	Diodia virginiana
Cardinal flower	Lobelia cardinalis
Carpet-weed	Mollugo verticillata
Cattail	Typha latifolia
Chervil, southern	Chaerophyllum tainturiera
Chickweed, common	Stellaria media
Chickweed, sticky mouse-eared	Cerastium glomeratum
Clearweed	Pilea pumila
Clover, white	Trifolium repens
Copperleaf, rhombic	Acalypha rhomboides
Craine's-bill, dove's foot	Geranium molle
Crownbeard, southern	Verbesina occidentalis
Cucumber-root, indian	Medeola virginiana
Daisy, common Peruvian	Gallinsoga quadriradiata
Dandelion	Taraxacum officinale
Dayflower	Commelina communis
Dayflower, Virginia	Commelina virginica
Daylily, orange	Hemerochallis fulva
Dead-nettle, purple	Lamium purpurem
Dock, curley	Rumex crispus
Dog fennel	Eupatorium capillifolium
Elephant's foot	Elephantopus carolinianus
False-nettle	Boehmeria cylindrica
Fireweed	Erechtites hieracifolia
Five fingers	Potentilla anadfensis
Five-fingers, running	Potentilla canadensis
Garlic, field	Allium vineale
Geranium, wild	Geranium maculatum

Goldenrod, tall	Solidago altissima
Groundsel	Senecio vulgaris
Heal all	Prunella vulgaris
Honewort	Cryptotaenia canadensis
Horse-nettle	Solanum carolinense
Horseweed	Conyza canadensis (Erigeron canadensis)
Indian pipes	Monotropa uniflora
Indian strawberry	Duchesnea indica
Ironweed	Vernonia noveboracensis
Jack-in-the-pulpit	Arisaema triphyllum
Jewelweed, orange	Impatiens capensis
Jumpweed	Persicaria virginiana (Tovara virginianum)
Lettuce, wild	Lactuca saligna
Lizard's tail	Saururus cernuus
Mayapple	Podophyllum peltatum
Mazus	Mazus pumilus (M. japonicus)
Monkey grass	Liriope muscari
Monkey-flower, Allegheny	Mimulus ringens
Nightshade, American	Solanum ptychanthum (S. americanum)
Orchid, crain-fly	Tipularia discolor
Pennyroyal, American	Hedeoma pulegioides
Pickerelweed	Pontederia cordata
Pimpernel, false	Lindernia dubia
Pinkweed	Persicaria pennsylvanica (Polygonum)
Pipsissewa	Chimaphila maculata
Plantain, American	Plantago rugelii
Polkweed	Phytolacca americana
Pondweed, common snailseed	Potamogeton diversifolius
Primrose, creeping	Ludwigia palustrus
Primrose, evening	Oenothera biennis
Primrose, winged water primrose	Ludwegia decurrens
Purslane, Milk	Chamaesyne maculata (Euphorbia supina)
Radish, wild	Raphanus raphanistrum
Ragweed	Ambrosia artemisiifolia
Ragweed	Ambrosia trifida
Rattlesnake-root, tall	Prenanthes altissima
Sida, arrowleaf	Sida rhombifolia
Smartweed, common	Persicaria hydropoper (Polygonum)
Smartweed, dense flowered	Persicaria densaflora (Poligonum)
Smartweed, long bristled	Persicaria longisetum (P. cespitosum)
Smartweed, swamp	Persicaria setacea (Polygonum)
Snakeroot, black	Sanicula canadensis
Solomon's seal	Polygonatum biflorum
Solomon's seal, false	Maianthemum racemosum (Smilacina)
Spanish needles	Bidens bipinnata

Speedwell	Veronica hederaefolia
Spiderwort, Asian	Murdannia keisak (Aneilema)
Spurge, eastern flowering	Euphorbia corrollata
St. John's wort, dwarf	Hypericum mutilum
Stitchwort, bog	Stellaria alsine
Sunflower, rough leaf	Helianthus strumosus
Tea, Mexican	Chenopodium ambrosioides
Thistle, yellow	Cirsium hordulum (Carduus spinosissimus)
Turnsole	Heliotropium indicum
Vervain, white	Verbena urticifolia
Violet, dooryard	Viola sororia (V. papilionacea)
Violet, southern wooly	Viola hirsutula
Water-willow	Ludwigia leptocarpa
Wild petuna, Carolina	Ruellia caroliniensis
Wild senna, northern	Senna hebecarpa
Willowweed	Persicaria lapathifolia (Polygonum)
Windflower	Anemonella thalictroides (Thalictrum)
Yellow wood-sorrel, common	Oxalis stricta

Ferns

Bracken fern	Pteridium aquilinum
Christmas fern	Polystichum acrostichoides
Ebony spleenwort	Asplenium platyneuron
Sensitive fern	Onoclea sensibilis
Grape fern	Botrychium biternatum
Southern lady fern	Athyrium asplenioides

Grasses, sedges & grass-like plants

Barnyard grass	Echinochloa crusgalli
Bulrush, woolgrass	Scirpus cyperinus
Bur-reed	Sparganium americanum
Cane, giant	Arundinaria gigantea
Cattail	Typha latifolia
Crabgrass	Digitaria sanguinalis
Crabgrass,smooth	Digitaria ischaemum
Cyperus	Cyperus echinatus (C. ovularis)
Dallis Grass	Paspalum dilatatum
Indian love grass	Eragrostis pilosa
Japanese grass	Microstegium vimineum
Johnson grass	Sorghum halepense
Monkey grass	Liriope muscari
Nut sedge, false	Cyperus strigosus
Panic grass, deer-tongue	Dichanthelium clandestinum (Panicum)
Panic Grass, fall	Panicum dichotomiflorum
Redtop	Tridens flavus
River oats	Chasmanthium latifolium (Uniola)
Rush, common	Juncus effusus
Rush, path	Juncus tenuis
Sedge	Carex crinita

Sedge	Carex frankii
Sedge	Carex vulpinoidea
Sedge, annual greenhead	Kyllinga pumila (Cyperus tenuifolius)
Sedge, hop	Carex lupulina
Sedge, round hedge	Cyperus echinatus
Short husk, common	Brachyelytrum erectum
Speargrass	Poa annua
Vernal grass, sweet	Anthoxanthum odoratum
White cut grass	Leersia virginica
Wild rye grass	Elymus virginicus
Woodreed, common	Cinna arundinaceae
Yardgrass	Eleusine indica

SHRUBS

Black willow	Salix nigra
Blackberry	Rubus argutus
Bladdernut	Staphylea trifolia
Buttonbush	Cephalanthus occidentalis
Cherry, weeping	Prunus sp.
Dewberry, northern	Rubus flagellaris
Dogwood, common	Cornus florida
Dogwood, silky	Cornus amomum
Elderberry, common	Sambucus canadensis
Euonymous, winged	Euonymus alatus
Hawthorne	Crataegus uniflora
Hazelnut	Corylus cornuta var. cornuta
Hazelnut, American	Corylus americana
Himalaya-berry	Rubus bifrons
Holly	Ilex opaca
Holly, Japanese	Ilex crenata
Hydrangia, snowy	Hydrangea radiata (H. arborescens var. radiata)
Laurel, Carolina cherry	Prunus caroliniana
Laurel, Japanese	Aucuba japonica
Laurel, monutain	Kalmia latifolia
Mahonia, Chinese	Mahonia bealei
Mistletoe	Phoradendron serotinum
Mulberry, red	Morus rubra
Mulberry, white	Morus alba
Nandina	Nandina domestica
Oregon grape	Mahonia bealei
Photinia, red-tip	Photinia serratifolia
Privet (Chinese)	Ligustrum sinense
Privet (Japanese)	Ligustrum japonicum
Privet, glossy	Ligustrum lucidum
Redbud	Cercis canadensis
Rose mallow	Hibiscus moscheutas
Rose, multiflora	Rosa multiflora
Rose-of-sharon	Hibiscus syriacus
Sassafras	Sassafras albidum
Silverbell, Carolina	Halesia carolina

Silverberry, autumn	Elaeagnus pungens
Smooth arrowwood	Viburnum recognitum (V. dentatum var. lucidulum)
Spring silverberry	Elaeagnus umbellata var. parvifolia (E. umbellata)
Strawberrybush	Euonymus americanus
Sweetshrub	Calacanthus floridus
Tag alder	Alnus serrulata

Vines

Ground cover and/or climbing vines

Anglepod, common	Gonolobus gonocarpa (Matelea)
Arrowvine	Polygonum sagittatum
Bindweed, hedge	Calystegia sepium
Buckwheat, crested climbing	Fallopia scandens (Polygonium)
Bur cucumber	Sicyos angulatus
Catbrier	Smilax bona-nox
Cinnamon vine	Dioscorea polystachya (D. batates)
Coralbeads	Cocculus carolinus
Cross-vine	Bignonia caprelolata (Anisostichus)
Dodder	Cuscuta gronovii
English ivy	Hedera hexix var. helix
Grape, possum	Vitis cinera var. baileyana (V. baileyana)
Greenbrier, bull	Smilax rotundifolia
Greenbrier, whiteleaf	Smilax glauca
Ground nut, common	Apios americana
Hemp, cliumbing	Mikania scandens
Honeysuckle, coral	Lonicera sempervirens
Honeysuckle, Japanese	Lonicera japonica
Hops, Japanese	Humulus japonicus
Hydrangia, climbing	Decumaria polystachyna (D. barbara)
Jessamine, Carolina	Gelsimium sempervirens
Kudzu	Pueraria montana var. lobata (P. lobata)
Leatherflower, northern	Clematis viorona
Maypops	Passiflora incarnata
Morning glory, common	Ipomoea purpurea
Muscadine	Vitis rotundifolia
Passion flower, yellow	Passiflora lutea
Poison ivy	Toxicodendron radicans var. radicans (Rhus radicans)
Spinypod, Carolina	Matelea carolinensis
Sweet potato, wild	Ipomoea pandurata
Trumpet creeper	Campsis radicans
Virginia creeper	Parthenocissus quinquefolia
Virgin's bower	Clematis virginiana
Wintercreeper	Euonymus fortunei
Wisteria	Wisteria florabunda
Yam, wild whorled	Dioscorea quaternata (D. villosa)

Vining non-climbing ground cover

Gill over the ground	Glechoma hederacea
Partridge berry	Mitchella repens
Periwinkle, greater	Vinca major

Trees

Ash, American	Fraxinus americana
Beech, American	Fagus grandifolia
Birch, river	Betula nigra
Boxelder	Acer negundo
Cherry, black	Prunus serotina
Cottonwood	Populus deltoides
Dogwood	Cornus florida
Elm, winged	Ulmus alata
Elm, slippery	Ulmus rubra
Fringe tree	Chionanthus virginicus
Hickory, bitternut	Carya cordiformis
Hickory, mockernut	Carya tomentosa
Holly, American	Ilex opaca
Ironwood	Carpinus caroliniana
Locust, honey	Gelditsia triacanthos
Magnolia, southern	Magnolia grandiflora
Maple, red	Acer rubrum
Maple, silver	Acer saccharinum
Mimosa	Albizia julibrissin
Mulberry, paper	Broussonetia papyrifera
Mulberry, red	Morus rubra
Oak, southern red	Quercus falcata
Oak, water	Quercus nigra
Oak, white	Quercus alba
Oak, red	Quercus rubra
Oak, willow	Quercus phellos
Paulownia, royal	Paulownia tomentosa
Pawpaw	Asimina triloba
Pine, loblolly	Pinus taeda
Pine, shortleaf	Pinus echinata
Pine, Virginia	Pinus virginiana
Pinus, white	Pinus strobus
Redbud	Cercus canadensis
Red cedar	Juniperus virginiana
Sassafras	Sassafras albidum
Serviceberry	Amelanchier arborea
Silverbell, Carolina	Celtis sp.
Sourwood	Oxydendron arboreum
Sugarberry	Juglans cinerea
Sweetgum	Liquidamber styraciflua
Tree of heaven	Ailanthus altissima
Tulip tree	Liriodendron tulipifera
Tupelo	Nyssa sylvatica
Walnut, black	Juglans nigra
Willow, black	Salix nigra

G.R. Davis Jr., professor of biology at Wofford College, has dabbled in photography since high school. He enjoys documentary and travel photography where he delights in the unexpected, the incongruous, and subjects that are not particularly beautiful. He finds settings of natural beauty (such as the Cottonwood Trail) challenging to capture on film. Rarely satisfied with his own photographs, he hopes persistence will lead to improvement.

Peter L. Schmunk is a frequent and avid traveler to natural and cultural sites. A professor of art history at Wofford College, he leads tours of European cities and regularly returns to the West, where he was raised, to backpack in mountain wilderness areas. Whether in Paris or North Carolina's Pisgah Forest, in Barcelona or his own backyard, he seeks to photograph the beauty the world has to offer and has recently exhibited his photographs at several Spartanburg area venues.

Thomas Webster is the father of three and has been an elementary school teacher for the past ten years. While at USC-Spartanburg, he was the fiction editor of *The Prickly Pear*. Thomas has worked as a freelance writer for *Creative Loafing* and various local papers, and his fiction and poetry have appeared in several publications. He is the recent recipient of a Hub City Writing Prize and a winner of the South Carolina Fiction Project.

My photographs are dedicated to my parents who, though not gamblers, risked losing their money when they purchased my first camera from a New York mail order company in response to my fervent pleading as a high school sophomore in 1972. My work is also dedicated to my wife who continues to encourage and support my interests in photography and travel.

— G.R. Davis Jr.

My photographs are dedicated to the memory of my father, who long ago introduced me to photography and to wild places and who was an interested, affirming viewer of the images I gathered.

— Peter L. Schmunk

This book is dedicated to my parents for setting me free to explore the woods, my wife for encouraging it still, and my children for giving me new reasons to marvel at and value Creation.

Were this book a record of my own knowledge, it would be a very short book indeed. I am indebted to the following for graciously sharing their expertise with me: Dr. Lyle Campbell, Sarah Campbell, Dr. Douglas Jensen, and Dr. Jack Turner. Each of these folks took walks on the trail with me, and I always came away wiser for having been in their company. Thank you to Dr. Gillian Newberry who put a great deal of effort into researching the Cottonwood's plants. She has put them on record in the University of South Carolina Upstate's Herbarium, one of Spartanburg's most overlooked treasures. Thanks, too, to all the helpful staff that populate the Spartanburg County Library. I appreciate the work of Mary Walter, SPACE, and all of the folks who volunteer their time to maintain the trail. Finally, my gratitude to the Hub City Writers Project and, especially, Betsy Teter. You will probably never realize just how much this opportunity has meant to me. Thank you.

— Thomas Webster

COLOPHON

The path leading to *Cottonwood Trail*, the second large-format, full-color Hub City coffeetable "nature book," finally emerged from the woods in late spring 2006 with four seasons' worth of 35mm slides to scan. A Nikon Super Coolscan 8000 with a five slide holder made quick work of the digital conversion. Several external FireWire hard drives provided safe storage and easy transportation from one PowerMac G3 to another. The text face is Bookman and the headline font is Tragan. This colorful first edition of 1000 is only available in hardback. A sad note on design programs: this could be the last, or the next to last, of Hub City's books to be designed in the venerable Pagemaker and "experienced" G3s. InDesign is Adobe's new industry standard and even stalwart, conservative Olencki Graphics will sooner or later have to abandon the reliable OS 9.xx for Tiger (OS X 10.4.6) and the "brave new digital world" of PDFs as the InDesign only runs on OS10.xx. A "new" suite of four G4s (many sporting dual processors) lay in wait at the designer's feet. Deadlines and a lack of experience are the only stumbling blocks in the way of utilizing their awesome power and speed. The almost daily indulgences of early morning fresh-roasted coffee from Little River Roasting Company and the noontime pick-me-ups of sweet tea and roast beef sandwiches from the 45-year-old Sugar-n-Spice fueled the creative spirits throughout the book's production.